100 square

The 100 number square has the first 100 numbers arranged in a 1...

There are lots of patterns to spot. The square is very useful for counting, addition and subtraction.

All the numbers ending in 1 are in the 1st column.

All the numbers ending in 2 are in the 2nd column and so on.

Pick any number. The number directly below it is 10 more.

46
56

When 10 is added to a number, only the tens digit changes.

1	2	3	4	5	6	7	8	9	10
11	12	13	14	15	16	17	18	19	20
21	22	23	24	25	26	27	28	29	30
31	32	33	34	35	36	37	38	39	40
41	42	43	44	45	46	47	48	49	50
51	52	53	54	55	56	57	58	59	60
61	62	63	64	65	66	67	68	69	70
71	72	73	74	75	76	77	78	79	80
81	82	83	84	85	86	87	88	89	90
91	92	93	94	95	96	97	98	99	100

Choose any number. The number directly above it is 10 fewer.

All the numbers ending in 0 (multiples of 10) are in the right hand column.

You can use the number square for addition.

38 + 5 = ?

Start on 38 and count on 5. When you reach the end of the row, carry on counting from the left of the row below, so:

38 + 5 = 43

To subtract count back, reversing the process!

Number lines

Look at the way solutions for number sentences using =, > and < can be illustrated on a number line.

8 + ( ) = 12   In an equality statement, there is a single solution for ( ), 4.

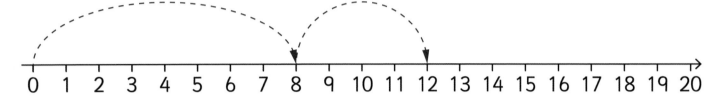

8 + ( ) > 12   With the > inequality statement, possible solutions for ( ) are 5 or any number greater than 5.

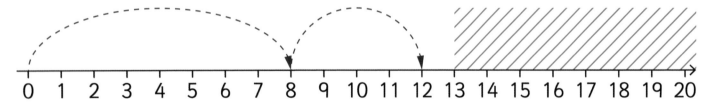

8 + ( ) < 12   With the < inequality statements possible solutions for ( ) are 0, 1, 2 or 3.

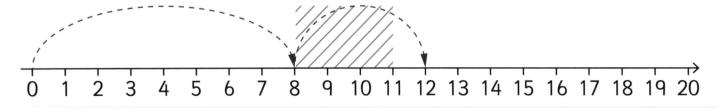

Quantity place value

... shows how many ones are in that place.

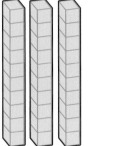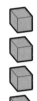

35 is made up of thirty and five.

The quantity of each digit is represented by 3 rods that show 10 each and 5 ones.

Column place value

... shows how many of those things that we have – **how many** ones in the ones column, **how many** tens in the tens column, **how many** hundreds in the hundreds column.

35 is made up of three 10s and five 1s.

The number of 10s and 1s is represented by 3 red 10s counters and 5 white 1s counters.

Names of parts of number sentences

addend	operation	addend	equals	answer
25	+	7	=	32
	add			total
	plus			altogether
				sum

minuend	operation	subtrahend	equals	answer
46	–	5	=	41
	subtract			difference
	take away			
	minus			

Partitioning

Splitting a number up into 10s and 1s to help with calculation.

Addition

35 + 24 = ☐

Partition both addends

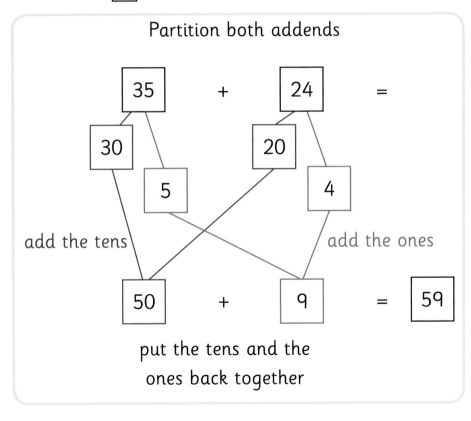

add the tens

add the ones

put the tens and the
ones back together

Partition one addend

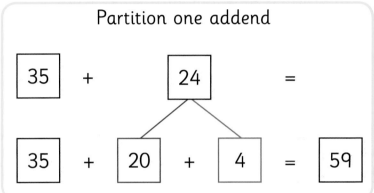

Subtraction

$35 - 24 = \square$

Partition both numbers

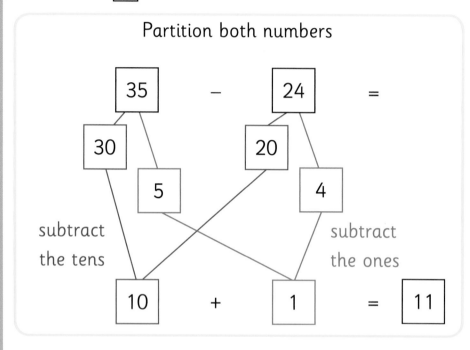

subtract the tens

subtract the ones

Partition only the subtrahend

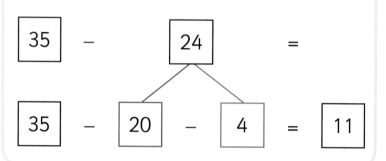

Compensation

Rounding a number close to a multiple of 10 and then compensating for that.

Addition

27 + 11 is the same as adding 10 and then adding another 1 (because 11 is 1 more than 10).

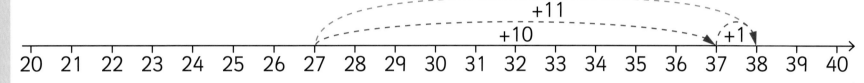

27 + 9 is the same as adding 10 and then subtracting 1 (because 9 is 1 less than 10).

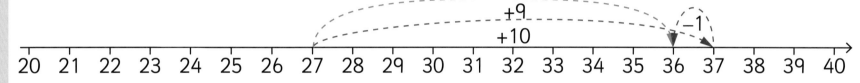

Subtraction

27 − 11 is the same as subtracting 10 and then subtracting another 1 (because 11 is 1 more than 10).

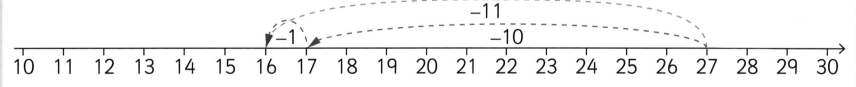

27 − 9 is the same as subtracting 10 and then adding 1 (because 9 is 1 less than 10).

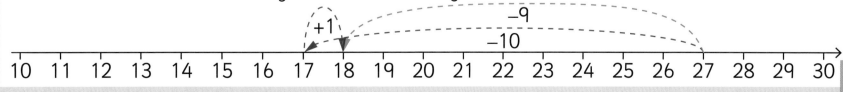

Re-ordering

Using the knowledge of the commutative law and changing the order of the numbers to make calculating more efficient. Re-ordering cannot be used for subtraction.

This looks quite hard.

24 + 7 + 16

This looks quite hard.

32 + 56

56 + 32

24 + 16 + 7 =

I can see a number bond to 10. If I add 24 and 16 first, this is much easier.

It's much easier to re-order and start from the largest number.

Bridging

Partitioning an addend or a subtrahend to bridge through a tens number to make calculation more efficient.

37 + 15 = 37 + 3 + 10 + 2

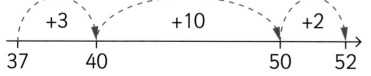

34 − 17 = 34 − 4 − 10 − 3

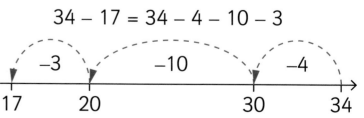

Regrouping

When there are not enough ones in the minuend to subtract the subtrahend, the minuend can be regrouped.

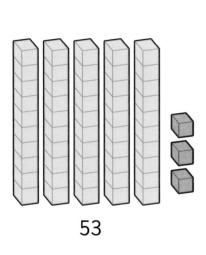

53

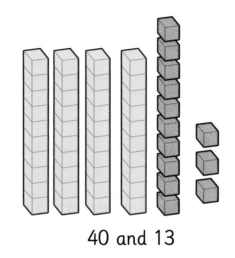

40 and 13

'Forty thirteen' is the same as 53, it has been regrouped.

$$\begin{array}{r} 4\;^{1} \\ \cancel{5}\;\;3 \\ -\;2\;\;9 \\ \hline \\ \hline \end{array}$$

Written column methods for calculation

$57 + 37 =$ $43 - 28 =$

1 Write the numbers vertically, so that the tens and the ones digits are in neat columns.

	5	7
+	3	7

	4	3
−	2	8

2 Look at the ones column first.

If it is an addition, add the ones together. If they make more than 10, carry 1 ten into the tens column.

If it is a subtraction, subtract the ones. If the top digit is smaller than the bottom digit, regroup the minuend to help.

	5	7
+	3	7
		4

1

	$^3\!4$	$^1 3$
−	2	8
		5

3 Look at the tens column next.

If it is an addition, add the tens together. Remember to include any tens you carried over.

If it is a subtraction, subtract the tens. Remember the minuend may have changed if you regrouped it.

	5	7
+	3	7
	9	4

1

	$^3\!4$	$^1 3$
−	2	8
	1	5

Year 2 children investigated the numbers of insects in their school wildlife garden.

They used tally marks to make a tally chart. Tally marks are drawn in sets of 5 lines, 4 vertical ones with the 5th as a diagonal line across them. They counted the tally marks to complete a frequency table.

Tally chart	
Ladybirds	ⅢⅢ |||
Beetles	ⅢⅢ	
Ants	ⅢⅢ ⅢⅢ	
Bees	||||

Frequency table	
Ladybirds	8
Beetles	6
Ants	11
Bees	4

From the data they drew a pictogram and a block diagram.

Pictogram

Block diagram

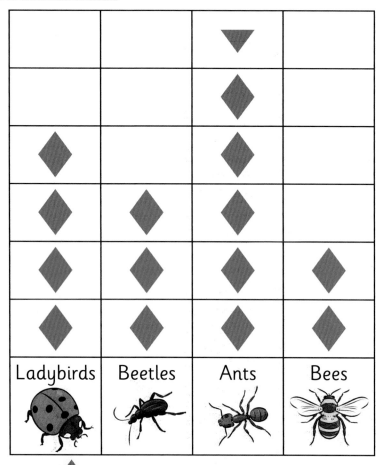

Each ◆ represents 2 insects (▼ is 1 insect)

The children also studied the trees in their wildlife area. Here is a block diagram of the trees they counted.

In this diagram each cell represents 1 tree.

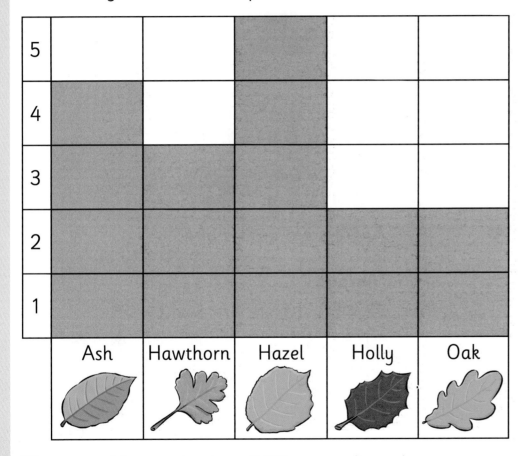

5					
4					
3					
2					
1					
	Ash	Hawthorn	Hazel	Holly	Oak

There are 16 trees in the wildlife area altogether.

4 + 3 + 5 + 2 + 2 = 16

In autumn they counted the number of apples on each apple tree in the orchard (before they ate them!)

This block diagram shows the number of apples on each apple tree.

Each cell in this diagram represents 5 apples.

50				
45				
40				
35				
30				
25				
20				
15				
10				
5				
	Apple tree A	Apple tree B	Apple tree C	Apple tree D

Multiplication

Equal groups

There are 5 groups of 2.

There are 6 groups of 4.

There are 3 flowers in each pot and there are 5 pots.

There are 15 flowers altogether.

This means 5 groups of 3.

Addition sentence: 3 + 3 + 3 + 3 + 3 = 15

Multiplication sentence: 5 × 3 = 15

Repeated addition on a number line

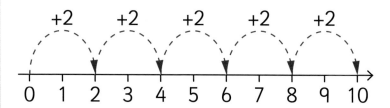

Arrays

2 × 5 = 10   5 × 2 = 10

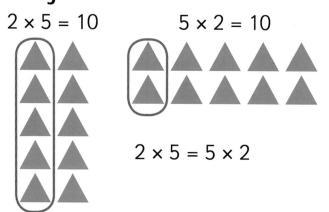

2 × 5 = 5 × 2

Twice, three times... ten times...

The number of 🌙 is twice the number of ⭐

The number of ➕ is 3 times the number of ▲

The number of ⚪ is 10 times the number of ⬟

 20    4

The number of spoons is 5 times the number of cups.

$4 \times 5 = 20$

The effect of multiplication and addition

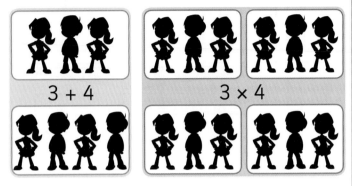

$3 + 4$ 3×4

Multiplying by 10

tens	ones
	7
7	0

$7 \times 10 = 70$

Equivalent multiplication sentences

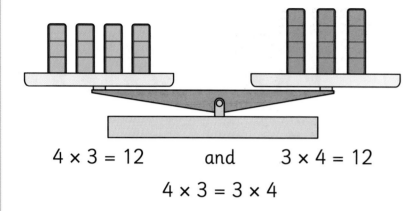
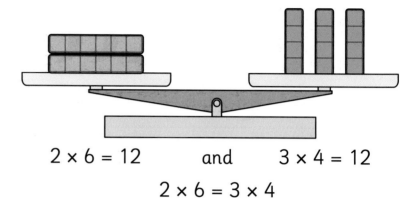

4 × 3 = 12   and   3 × 4 = 12

4 × 3 = 3 × 4

2 × 6 = 12   and   3 × 4 = 12

2 × 6 = 3 × 4

Division

Using grouping for repeated subtraction

Five groups of three can be made with the ⭐ in the diagram.

15 − 3 − 3 − 3 − 3 − 3 = 0

Repeated subtraction on a number line

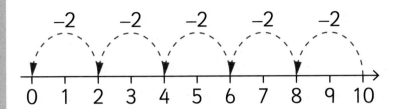

Dividing into '☐ equal groups' and 'groups of ☐'

Divide the 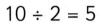 into two equal groups.

6 ÷ 2 = 3

Divide the ▲ into groups of 3.

6 ÷ 3 = 2

Arrays

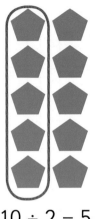   

10 ÷ 2 = 5   10 ÷ 5 = 2

Bar model

Represents 4 × 5 = 20 and 20 ÷ 4 = 5

Represents 7 × 6 = 42 and 42 ÷ 7 = 6

17

Part–whole relationships

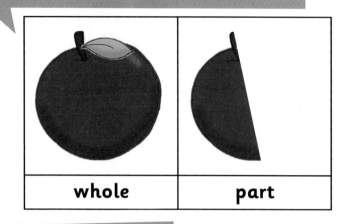

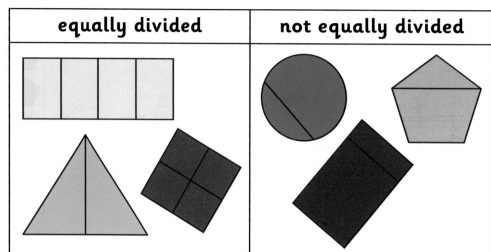

Unit fractions

$\frac{1}{2}$	
$\frac{1}{3}$	
$\frac{1}{4}$	
$\frac{1}{5}$	
$\frac{1}{6}$	

Quarters

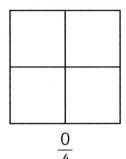 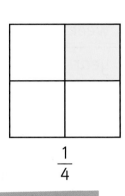 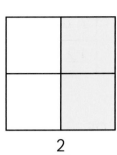 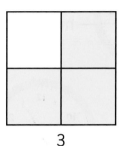 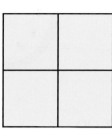

$$\frac{0}{4} \qquad \frac{1}{4} \qquad \frac{2}{4} \qquad \frac{3}{4} \qquad \frac{4}{4}$$

Fraction number lines

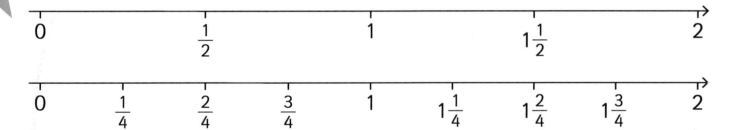

$$0 \qquad \frac{1}{2} \qquad 1 \qquad 1\frac{1}{2} \qquad 2$$

$$0 \qquad \frac{1}{4} \qquad \frac{2}{4} \qquad \frac{3}{4} \qquad 1 \qquad 1\frac{1}{4} \qquad 1\frac{2}{4} \qquad 1\frac{3}{4} \qquad 2$$

Fractions of quantities

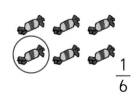 $\dfrac{1}{6}$ 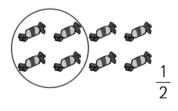 $\dfrac{1}{2}$ 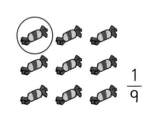 $\dfrac{1}{9}$  $\dfrac{1}{5}$

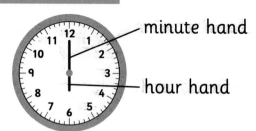
minute hand

hour hand

60 minutes	1 hour
24 hours	1 day
7 days	1 week
12 months	1 year

 7:00
7 o'clock

 7:05
5 past 7

 7:10
10 past 7

 7:15
$\frac{1}{4}$ past 7

 7:20
20 past 7

 7:25
25 past 7

 7:30
$\frac{1}{2}$ past 7

 7:35
35 minutes past 7
25 to 8

 7:40
40 minutes past 7
20 to 8

 7:45
45 minutes past 7
$\frac{1}{4}$ to 8

 7:50
50 minutes past 7
10 to 8

 7:55
55 minutes past 7
5 to 8

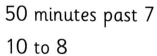

2018 Calendar

January 2018
S	M	T	W	T	F	S
	1	2	3	4	5	6
7	8	9	10	11	12	13
14	15	16	17	18	19	20
21	22	23	24	25	26	27
28	29	30	31			

February 2018
S	M	T	W	T	F	S
				1	2	3
4	5	6	7	8	9	10
11	12	13	14	15	16	17
18	19	20	21	22	23	24
25	26	27	28			

March 2018
S	M	T	W	T	F	S
				1	2	3
4	5	6	7	8	9	10
11	12	13	14	15	16	17
18	19	20	21	22	23	24
25	26	27	28	29	30	31

April 2018
S	M	T	W	T	F	S
1	2	3	4	5	6	7
8	9	10	11	12	13	14
15	16	17	18	19	20	21
22	23	24	25	26	27	28
29	30					

May 2018
S	M	T	W	T	F	S
		1	2	3	4	5
6	7	8	9	10	11	12
13	14	15	16	17	18	19
20	21	22	23	24	25	26
27	28	29	30	31		

June 2018
S	M	T	W	T	F	S
					1	2
3	4	5	6	7	8	9
10	11	12	13	14	15	16
17	18	19	20	21	22	23
24	25	26	27	28	29	30

July 2018
S	M	T	W	T	F	S
1	2	3	4	5	6	7
8	9	10	11	12	13	14
15	16	17	18	19	20	21
22	23	24	25	26	27	28
29	30	31				

August 2018
S	M	T	W	T	F	S
			1	2	3	4
5	6	7	8	9	10	11
12	13	14	15	16	17	18
19	20	21	22	23	24	25
26	27	28	29	30	31	

September 2018
S	M	T	W	T	F	S
						1
2	3	4	5	6	7	8
9	10	11	12	13	14	15
16	17	18	19	20	21	22
23	24	25	26	27	28	29
30						

October 2018
S	M	T	W	T	F	S
	1	2	3	4	5	6
7	8	9	10	11	12	13
14	15	16	17	18	19	20
21	22	23	24	25	26	27
28	29	30	31			

November 2018
S	M	T	W	T	F	S
				1	2	3
4	5	6	7	8	9	10
11	12	13	14	15	16	17
18	19	20	21	22	23	24
25	26	27	28	29	30	

December 2018
S	M	T	W	T	F	S
						1
2	3	4	5	6	7	8
9	10	11	12	13	14	15
16	17	18	19	20	21	22
23	24	25	26	27	28	29
30	31					

Mass

Mass is how heavy something is. We measure mass in grams (g) and kilograms (kg).

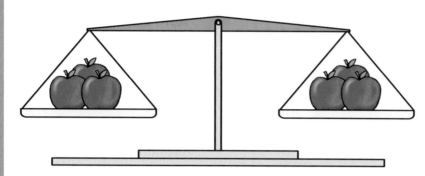

The pans are balanced, so the masses are equal.

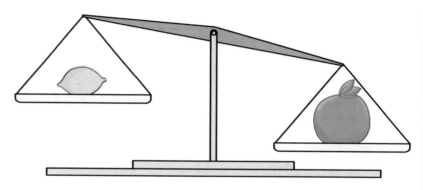

The orange is heavier than the lemon.

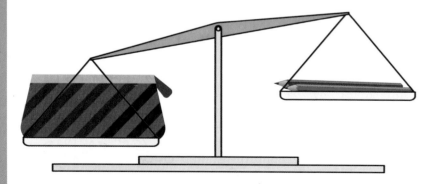

The pencils are lighter than the pencil case.

The mass of this banana is 95g.

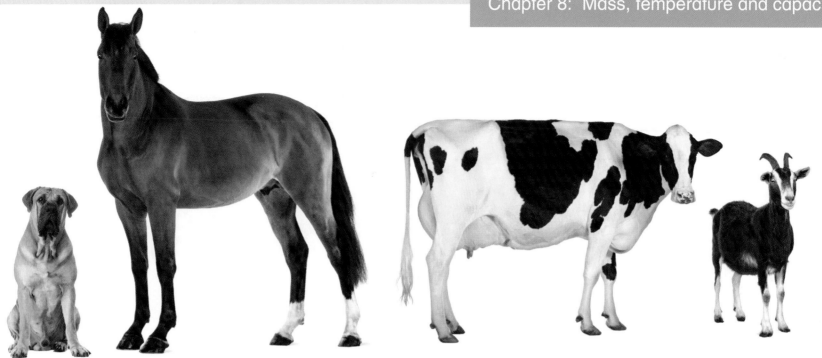

The mass of these animals would be measured in kilograms.

The mass of these animals would be measured in grams.

Temperature

Temperature is how hot or cold something is. We can measure temperature using a thermometer.

The unit we use to measure temperature is degrees Celsius (°C).

Water freezes when it is 0°C.

Water boils when it is 100°C.

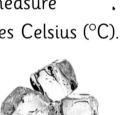

Our body temperature is 37°C.

This picture shows a place where it is very hot.

This picture shows a place where it is very cold.

Capacity and volume

Capacity is the amount a container can hold.

Volume is the amount of liquid in the container.

We measure capacity and volume in millilitres (ml) and litres (l).

We would measure this cough mixture in millilitres (ml).

| Empty | Half full | Full |

We would measure the water in this paddling pool in litres (l).

The bath has the greatest capacity; the mug has the least.

We use measuring jugs like this to measure the volume of liquids.

Position, turn and direction

The hands of a clock move in a clockwise direction:

The opposite to clockwise is anticlockwise.

The four main points of a compass are North, East, South and West.

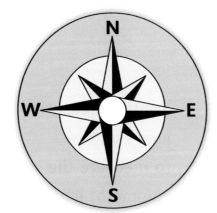

A quarter-turn clockwise:

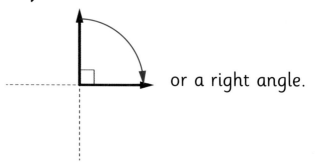 or a right angle.

A half turn anticlockwise:

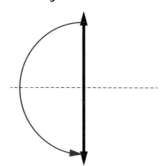

A three-quarter turn clockwise:

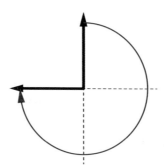

Symmetry

Symmetry is where one half of a shape is the mirror image of the other half. If you fold the shape and both halves fit on top of each other exactly, the shape is symmetrical.

Symmetrical

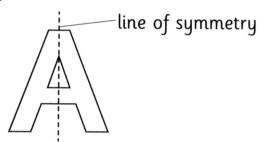
line of symmetry

One half is the mirror image of the other.

Asymmetrical

The two halves are different.

Polygons

Any 2-D shape with three or more straight sides belongs to a family of shapes called polygons. Here are some examples:

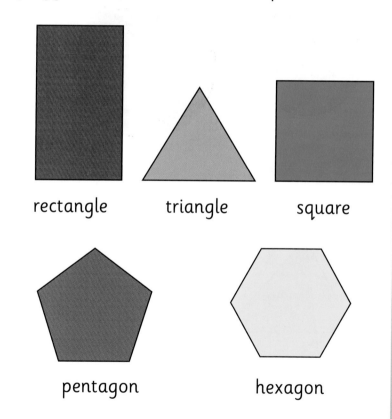

rectangle   triangle   square

pentagon   hexagon

Circle

A circle is a 2-D shape. It is not a polygon because it has curved sides.

3-D Shapes

Here are some examples of 3-D shapes.

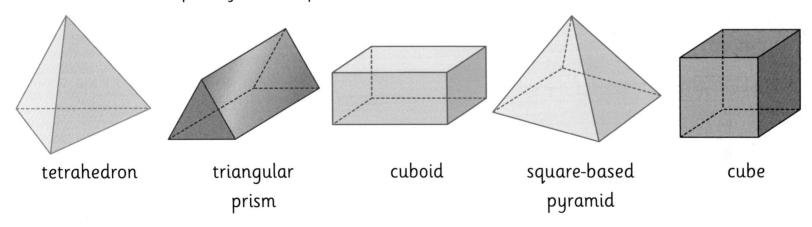

| tetrahedron | triangular prism | cuboid | square-based pyramid | cube |

100 square:

1	2	3	4	5	6	7	8	9	10
11	12	13	14	15	16	17	18	19	20
21	22	23	24	25	26	27	28	29	30
31	32	33	34	35	36	37	38	39	40
41	42	43	44	45	46	47	48	49	50
51	52	53	54	55	56	57	58	59	60
61	62	63	64	65	66	67	68	69	70
71	72	73	74	75	76	77	78	79	80
81	82	83	84	85	86	87	88	89	90
91	92	93	94	95	96	97	98	99	100

add: increase one number by another or put two numbers together

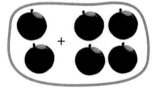

addend: the number being added, or added to, in an addition calculation, addend + addend = sum

$$4 + 3 = 7$$
↑ ↑
addend

addition: Join or put together two or more numbers or amounts. The symbol for addition is +

align: in the column method for calculation this means line up columns under each other, ones under ones and tens under tens

$$\begin{array}{r} 5\ 7 \\ +\ 3\ 7 \\ \hline 9\ 4 \end{array}$$

altogether: all, everything

everything

analogue time: time shown using hands on a clock

angle: the amount of turn between two straight lines that meet at a point

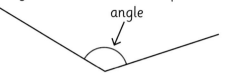

angle

anticlockwise: in the opposite direction to the way the hands of a clock move

array: a regular pattern of rows and columns for arranging symbols or objects

asymmetrical: does not have symmetry – one half is not a reflection of the other

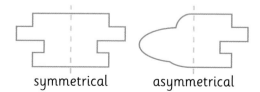

symmetrical   asymmetrical

29

balance: one side is the same as the other in some way

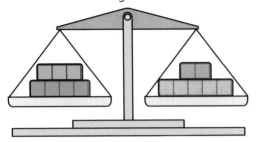

bar model: a diagram to show part–whole relationships

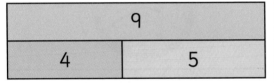

bead string: a string of beads, usually 10, 20, 50 or 100 used to support calculations

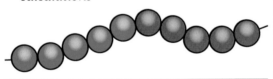

bigger: greater/larger than another quantity, object or number

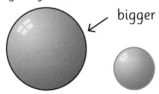
bigger

block diagram: a diagram where information is shown in blocks or as coloured squares

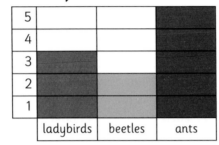

5			
4			
3			
2			
1			
	ladybirds	beetles	ants

boiling point: the temperature at which water boils, 100 degrees Celsius

bridging through 10: splitting up the number you are adding on or subtracting to reach a 10

+2 +3
0 1 2 3 4 5 6 7 8 9 10 11 12 13
8 + 5 = 13

−2 −3
0 1 2 3 4 5 6 7 8 9 10 11 12 13
13 − 5 = 8

calculate: work out the answer to a question about numbers

calculating: working out the answer to a question about numbers

calculation: a question about numbers written as a number sentence

$$3 + 6 = 9, \ 7 - 2 = 5$$

capacity: the amount that a container can hold

cell: a single rectangle or square in a block diagram

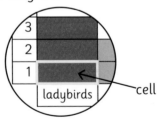
cell
ladybirds

check: work something out again to make sure it makes sense and is correct

circle: a closed 2-D shape with one curved outline consisting of points that are the same distance from the centre

clock: a device for measuring time

clockwise: in the direction that the hands of a clock move

coin: flat metal disc used as money, stamped with its value

cold: a low temperature

column: a vertical arrangement of numbers or objects, or a shaded vertical block

$$\begin{array}{r} 5\;7 \\ +\;3\;7 \\ \hline 9\;4 \end{array}$$

commutative: law for addition and multiplication that means the numbers can be swapped around without changing the answer

5 + 3 = 8 is the same as 3 + 5 = 8

compare: look at two or more things and see how they are the same or different

compensation: a mental calculation strategy in which a number is rounded to the nearest 10 to make the calculation easier, and the amount rounded up or down is compensated for at the end, for example 34 + 19, (34 + 20) − 1

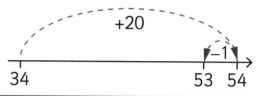

cone: a 3-D object with a circular base joined to a curved side that ends in a point, like an ice-cream cone

count: read or think the number names in order, saying one number name for each object

count all: count everything to find the total

count back: start counting part-way through the counting sequence, continuing by repeatedly counting one less or fewer

count on: start counting partway through the counting sequence, continuing by repeatedly counting one more

cube: a 3-D shape with 6 square faces

cuboid: a 3-D shape with 6 rectangular faces – the opposite faces are the same

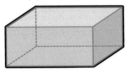

cylinder: a 3-D shape with two circular ends joined by a curved surface

data: information

Frequency table	
Ladybirds	8
Beetles	6
Ants	11

day: a period of time that is 24 hours

degrees Celsius: unit for measuring temperature

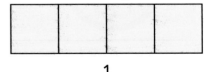

denominator: the number of equal parts something has been divided into

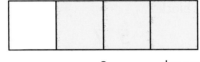

$\dfrac{3}{4}$ ← denominator

diagonal (line): a straight line that is slanting

difference: the amount or quantity by which one thing is different to another

   the difference between these 2 sticks of cubes is 3 cubes

digit: any of the symbols 0, 1, 2, 3, 4, 5, 6, 7, 8, 9

digital time: time written in numbers, for example, 11:30

divide: splitting a whole into equal parts or sharing into equal parts

$\dfrac{3}{4}$

dividend: the whole before it is divided

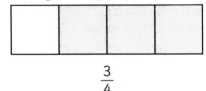

1

division: the act of dividing

divisor: the number that you divide by

$$6 \div \overset{\text{divisor}}{\textcircled{3}} = 2$$

double: a number becomes twice as large when it is doubled

Double 3 is 3 + 3, which is 6

duration: how long something lasts

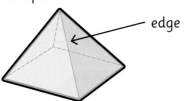

Swimming at 3 o'clock for 1 hour ← duration

edge: a line where two surfaces of a solid shape meet

edge

empty: contains nothing

equal: the same quantity, size or value

10   10

equals: Has the same value. The symbol for equals is =

$$2 + 2 = 4$$

equivalent: equal in quantity, size or value

$$\frac{1}{2} = \frac{2}{4} \qquad 3 + 4 = 2 + 5$$

even: a number that can be divided by 2 with nothing left over

    10

exchange: change 1 ten for 10 ones, or 10 ones for 1 ten

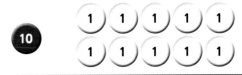

face: a flat surface of a 3-D shape

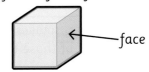 face

faces: flat surfaces of a 3-D shape

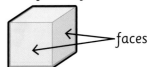 faces

fact family: all the addition and subtraction calculations which can be made with one set of three numbers in a part-whole relationship

$$2 + 3 = 5$$
$$5 - 3 = 2$$
$$5 - 2 = 3$$

factor: a number or quantity that when multiplied with another produces a given number

$$2 \times 3 = 6 \qquad 1 \times 6 = 6$$
factors of 6 are 1, 2, 3, 6

33

factor pair: A product can be made by multiplying a pair of factors together. This is a factor pair. Most products have more than one factor pair.

factors pairs of 12
1 × 12, 2 × 6, 3 × 4

fewer, fewest: smaller, smallest number or smallest quantity of those being compared

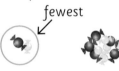

fewest

freezing point: when water begins to freeze, zero degrees Celsius

frequency table: a table that shows how many times a certain result occurs

Frequency table	
Ladybirds	8
Beetles	6
Ants	11

full: all space is filled up, nothing more can be fitted in

gram: a unit for measuring mass

greater: larger, more

greater

greater than (>): bigger in size or number

5 > 2

greatest: largest, most, biggest

 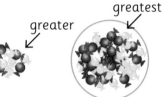

greater

greatest

group: a number of objects or images grouped together

half: one of two equal parts of the whole

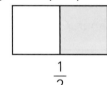

$\frac{1}{2}$

half an hour: a period of time that is 30 minutes in length

half full: half the space is filled up

half turn: A turn equal to two right angles. After a half turn you end up facing in the opposite direction that you started from.

hands: moving pointers on an analogue clock which show the hours and minutes

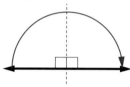

hands

heavier: has a greater mass than another object

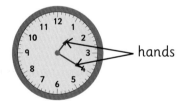

heavier

heaviest: has the greatest mass of a number of objects

heaviest

heavy: has a great mass

heptagon: a 2-D shape with 7 straight sides and 7 vertices

hexagon: a 2-D shape with 6 straight sides and 6 vertices

hot: a high temperature

hour: a period of time that is 60 minutes in length

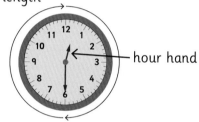

hour hand

inequality symbol: used to show that two amounts are not equal in value, < is less than, > is greater than

input: when used with a function, it shows the quantity going into the machine

input

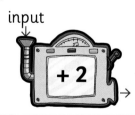

interval: a period of time between two events

It started raining at 2 o'clock and stopped at 3 o'clock.

---

inverse: An operation that is opposite to another. It undoes what has been done.

$$12 - 4 = 8 \qquad 8 + 4 = 12$$

---

is greater than: used to indicate the larger amount when comparing two quantities, symbol >

---

is less than: used to indicate the smaller amount when comparing two quantities, symbol <

---

kilogram: a unit for measuring mass

---

known: values which can be stated

 6 apples

---

largest: the biggest in size

largest

---

least: the smallest in quantity

least

---

less: smaller in quantity

less

---

less than (<): smaller in size or number

2 < 5

---

light: has a small mass

light   lighter   lightest

---

lighter: has a smaller mass than another object

light   lighter

---

lightest: the smallest mass of a number of objects

light   lighter   lightest

litre: a unit for measuring capacity and volume

mass: how heavy something is

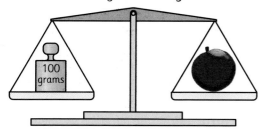

method: how to do something

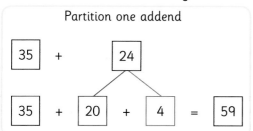

millilitre: a unit for measuring capacity and volume

minuend: The whole, the number being subtracted from.

minuend – subtrahend = difference

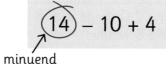

minuend

minute: A short period of time. There are 60 minutes in an hour.

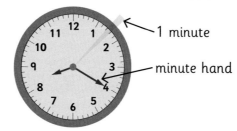

1 minute

minute hand

model: a representation of something else

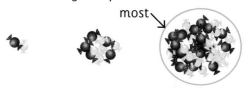

money: coins and banknotes

month: A period of time made up of at least 28 days. There are 12 months in a year.

May 2017						
S	M	T	W	T	F	S
	1	2	3	4	5	6
7	8	9	10	11	12	13
14	15	16	17	18	19	20
21	22	23	24	25	26	27
28	29	30	31			

most: the biggest quantity or number of those being compared

most

multiple: when you multiply one whole number by another, the result is a multiple of the starting number

$2 \times 3 = 6$
6 is a multiple of 2 and 3

multiply and **multiplication:** combining equal quantities

$$2 + 2 + 2 + 2 + 2 + 2 = 12$$
$$2 \times \mathbf{6} = 12$$

not equal: not the same quantity, size or value

number bond: a way of showing how two numbers combine to make another number, for example 6 and 4 combine to make 10

number line: a line of numbers in order, equally spaced and increasing in value from left to right towards infinity

0 1 2 3 4 5 6 7 8 9 10

number sense: ability to work flexibly with numbers and quantities; understanding of different representations and place value

number sentence: a mathematical sentence written with numerals and symbols

$$13 + 7 + 6 = 26$$

numerator: the number of parts we have

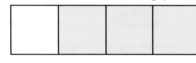

$\frac{3}{4}$ ⟵ numerator

octagon: a 2-D shape with 8 straight sides and 8 vertices

odd: numbers that will always have 1 left over when divided by 2

one fifth: one of five equal parts of the whole

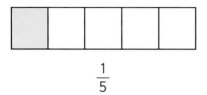

$\frac{1}{5}$

one quarter: the name of each of the parts that are made when something is split into 4 equal parts

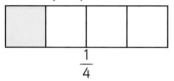

$\frac{1}{4}$

one third: the name of each of the parts that are made when something is split into 3 equal parts

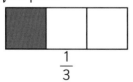

$\frac{1}{3}$

one(s): in a two-digit number, the second digit shows the number of ones

67 ⟵ ones

operation: a mathematical process, often +, −, x or ÷

part: a portion or segment of a whole

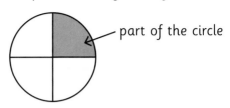

part of the circle

part–whole model: a diagram to show the relationship between two or more parts and the whole that they are parts of

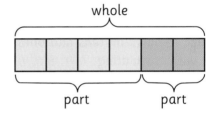

partition: split a number into two or more parts (often into 10s and 1s)

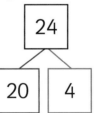

partitioning tree: a diagram to show how a number has been partitioned

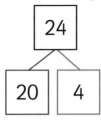

pattern: a regular arrangement of numbers or objects which follow a rule

pentagon: a 2-D shape with 5 straight sides and 5 vertices

pictogram: A diagram to show information. A picture is used to represent a number of items.

| Monday | Tuesday | Wednesday |

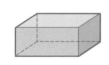 = 1 sweet    = 2 sweets

place value: The value of a digit depends on its place or position in the number. The column a digit is in tells us its value. The '3' in 35 is equal to 3 tens.

polygon: a 2-D shape with three or more straight sides

pound: a unit of money in the UK

prism: a 3-D shape with two identical ends and flat faces

product: the answer when two numbers are multiplied together

$2 \times 6 = 12$ ← product

pyramid: a 3-D shape with a polygon base and triangular faces which meet at the top (apex)

quadrilateral: a 2-D shape with 4 straight sides and 4 vertices

quantity: a number of objects or an amount of something

quarter: one of four equal parts of the whole

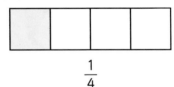

$$\frac{1}{4}$$

quarter of an hour: a period of time that is 15 minutes in length

quarter turn: a turn which is one quarter of a whole turn, also called a right angle

quotient: when a number is divided by another number, the answer is the quotient

$12 \div 2 = ⑥$ ← quotient

reasoning: thinking about what you know to help you work out something you do not yet know

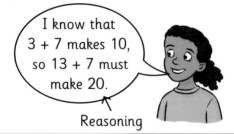

I know that 3 + 7 makes 10, so 13 + 7 must make 20.

Reasoning

re-combine: put numbers back together after partitioning

rectangle: A 2-D shape with 4 sides and 4 vertices. Opposite sides are equal in length.

reflection: The mirror image of something. One side matches the other.

mirror line

reflection   original

regroup: re-partition tens and ones to help with calculating

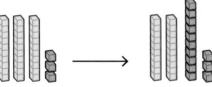

3 tens and 3 ones can be regrouped into 2 tens and 13 ones to make subtraction easier

relationship: how one number or object is related to another

re-order: put numbers in a different order to help with calculating

This looks quite hard.

32 + 56

56 + 32

It's much easier to re-order and start from the largest number.

---

repeated addition: adding the same value repeatedly

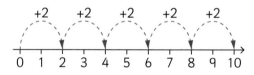

---

repeated subtraction: subtracting the same value repeatedly

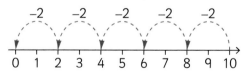

---

repeating: the same thing happening again

---

reverse: swap two numbers over, so the second thing becomes the first thing and the first thing becomes the second thing

32 + 56

56 + 32

---

right angle: a quarter turn

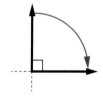

---

rotation: turning an object around a fixed point

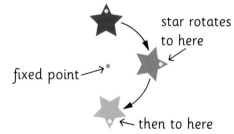

star rotates to here

fixed point →

← then to here

---

row: an arrangement of numbers or objects from side to side (horizontal)

1 2 3 4 5 6 7 8 9 10 11 12

---

scale: a straight line with equally spaced markings

---

segment: a portion or section of a whole

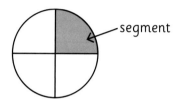

← segment

---

sequence: a set of things that follow a rule or pattern

1, 2, 4, 1, 2, 4, 1, 2, 4, 1, 2, 4

---

series: a set of things linked in some regular way

2, 4, 6, 8, 10, 12, 14, 16, 18

---

side: the line joining two vertices of a 2-D shape

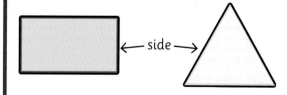

← side →

slice: a portion or section of a whole

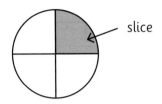

slice

smaller: less than another quantity, object or number

5 is a smaller number than 8

smallest: the least quantity, object or number

3 is the smallest number here

8 4 3 11 9 15

solution: the answer to a problem

12 ÷ 2 = ⑥ ← solution

sort: group numbers and things together according to their features

odd numbers	even numbers
3 1 5 7 9	4 2 8 6

sphere: a 3-D shape that is perfectly round like a ball

spider diagram: lines radiating out from a central core with a value in the core and at the end of each line

18 16 27 34 22 15 13 12 31

square: A 2-D shape with 4 sides and 4 vertices. All sides are the same length.

square-based pyramid: a 3-D shape with a square base and four triangular faces that meet at an apex

statistical table: a table containing data

Frequency table	
Ladybirds	8
Beetles	6
Ants	11

statistics: numerical facts that give information about something

Frequency table	
Ladybirds	⑧
Beetles	⑥ ← statistics
Ants	⑪

subitise: know how many without counting

subtraction: finding the difference between two numbers

12 − 2 = ⑩ ← difference

subtrahend: The number being subtracted from the minuend.
minuend − subtrahend = difference

sum: total, the answer when all addends have been added

12 + 2 = ⑭ ← sum

survey: a way of collecting data by asking people questions

symbol: a mark with a particular meaning, for example +, −, =, <, >

symmetrical: having symmetry

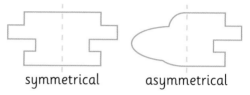

symmetrical   asymmetrical

symmetry: a shape has symmetry when one half is a mirror image of the other half

mirror line

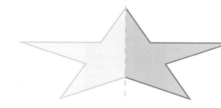

table: An arrangement of numbers in rows and columns. Tables set out information so that it is easy to find things out.

Input	Output
1	3
2	4
3	5
4	6

take away: a calculation to find the difference between two numbers

12 − 2 = ⑩ ← difference

tally: a way to count things in sets of 5 using lines to represent numbers

tally chart: a chart that uses tally marks to show how many of each thing

Tally chart	
Ladybirds	ЖⱠ |||
Beetles	ЖⱠ	
Ants	ЖⱠ ЖⱠ	

ten family: numbers with the same value in the tens place

10, 11, 12, 13, 14, 15, 16, 17, 18, 19

ten frame: a 5 by 2 grid

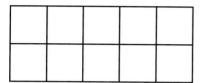

ten(s): in a two-digit number, the first digit shows the number of tens

tens →

tetrahedron: a 3-D shape with a triangular base and three triangular sides that meet at a point, also called a triangular-based pyramid

three dimensional (3-D): A shape is three dimensional if it has length, width and height. A cube is an example of a 3-D shape.

three-quarters of an hour: a period of time that is 45 minutes in length

times: another name for multiply

$$2 \times 6 = 12$$

total: another word for sum, the result of addition

$$12 + 2 = \textcircled{14} \longleftarrow \text{total}$$

two-digit number: number with digits in the tens and in the ones place

tens ones
3 5

unit fractions: a fraction where the top number is 1

$$\frac{1}{4}$$

unknown: A value which is not known but can be worked out. There may be more than one possible value.

$$10 + \boxed{?} = 12$$

value: another word for quantity

vertex: a point where two or more straight lines meet

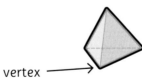

vertex $\longrightarrow$

vertical: In an up and down direction. A table leg is vertical.

vertices: more than one vertex

vertices

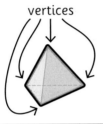

volume: the amount in a container

1 litre

warm: a temperature between cold and hot

week: a period of time made up of 7 days

May 2017						
S	M	T	W	T	F	S
	1	2	3	4	5	6
7	8	9	10	11	12	13
14	15	16	17	18	19	20
21	22	23	24	25	26	27
28	29	30	31			

weigh: measure how heavy something is

whole: sum of the parts

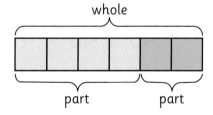

word problem: a mathematical question written in words

Max has 7 sweets and four friends. He gives 1 sweet to each friend and has 1 himself. How many sweets does Max have left?

year: a period of time made up of 12 months

1	2	3	4	5	6	7	8	9	10
11	12	13	14	15	16	17	18	19	20
21	22	23	24	25	26	27	28	29	30
31	32	33	34	35	36	37	38	39	40
41	42	43	44	45	46	47	48	49	50
51	52	53	54	55	56	57	58	59	60
61	62	63	64	65	66	67	68	69	70
71	72	73	74	75	76	77	78	79	80
81	82	83	84	85	86	87	88	89	90
91	92	93	94	95	96	97	98	99	100

William Collins' dream of knowledge for all began with the publication of his first book in 1819.

A self-educated mill worker, he not only enriched millions of lives, but also founded a flourishing publishing house. Today, staying true to this spirit, Collins books are packed with inspiration, innovation and practical expertise. They place you at the centre of a world of possibility and give you exactly what you need to explore it.

Collins. Freedom to teach.

Published by Collins
An imprint of HarperCollinsPublishers
The News Building
1 London Bridge Street
London
SE1 9GF

Browse the complete Collins catalogue at
www.collins.co.uk

© HarperCollinsPublishers Limited 2017

10 9 8 7 6 5 4 3

978-0-00-822596-4

Learning Books Series Editor: Amanda Simpson
Practice Books Series Editor: Professor Lianghuo Fan
Authors: Laura Clarke, Caroline Clissold, Sarah Eaton, Linda Glithro, Jane Jones, Steph King, Brian Macdonald, Cherri Moseley, Paul Wrangles.

British Library Cataloguing in Publication Data
A catalogue record for this publication is available from the British Library.

Publishing Manager: Fiona McGlade
In-house Editor: Nina Smith
In-house Editorial Assistant: August Stevens
Project Manager: Emily Hooton
Copy Editors: Tracy Thomas, Catherine Dakin
Proofreader: Mairi Sutherland
Cover design: Kevin Robbins and East China Normal University Press Ltd
Cover artwork: Daniela Geremia
Internal design: Amparo Barrera
Typesetting: Ken Vail Graphic Design and 2Hoots Publishing Services Ltd
Illustrations: Matt Ward (Beehive Illustration)
Production: Rachel Weaver

Printed in Great Britain by Martins the Printers

Photo acknowledgements
The publishers wish to thank the following for permission to reproduce photographs. Every effort has been made to trace copyright holders and to obtain their permission for the use of copyright materials. The publishers will gladly receive any information enabling them to rectify any error or omission at the first opportunity.

(t = top, c = centre, b = bottom, r = right, l = left)
p22 br (banana) bergamont/Shutterstock, p22 br (scale) Fedorov Ivan Sergeevich/Shutterstock, p23 tl Pavel Shlykov/Shutterstock, p23 tcl Eric Isselee/Shutterstock, p23 tcr Eric Isselee/Shutterstock, p23 tr Eric Isselee/Shutterstock, p23 bl Marco Uliana/Shutterstock, p23 bcl Eric Isselee/Shutterstock, p23 bcr YK/Shutterstock, p23 br Galushko Sergey/Shutterstock, p24 tl Lipskiy/Shutterstock, p24 tcl Erik Svoboda/Shutterstock, p24 tcr Arvind Balaraman/Shutterstock, p24 tr Kinga/Shutterstock, p24 cl Valentyn Volkov/Shutterstock, p24 bl Petr Malyshev/Shutterstock, p24 bc f9photos/Shutterstock, p24 br Li Hui Chen/Shutterstock, p25 tl Andrey_Kuzmin/Shutterstock, p25 bl Ttatty/Shutterstock, p25 bcl koosen/Shutterstock, p25 bcr flowerstock/Shutterstock, p25 tr Denis Tabler/Shutterstock, p25 cr sumroeng chinnapan/Shutterstock, p25 br M. Unal Ozmen/Shutterstock, p30 Petr Malyshev/Shutterstock, p31 Valentyn Volkov/Shutterstock, p32 Lipskiy/Shutterstock, p34 Erik Svoboda/Shutterstock, p35 Erik Svoboda/Shutterstock, p37 Denis Tabler/Shutterstock, p45 tl Erik Svoboda/Shutterstock, p45 tc (banana) bergamont/Shutterstock, p45 tc (scale) Fedorov Ivan Sergeevich/Shutterstock.